JAN. 10

Test Results for Forensic Media
Preparation Tool:
Darik's Boot and Nuke 1.0.7

NCJ 228983

Kristina Rose
Acting Director, National Institute of Justice

This report was prepared for the National Institute of Justice, U.S. Department of Justice, by the Office of Law Enforcement Standards of the National Institute of Standards and Technology under Interagency Agreement 2003-IJ-R-029.

The National Institute of Justice is a component of the Office of Justice Programs, which also includes the Bureau of Justice Assistance, the Bureau of Justice Statistics, the Office of Juvenile Justice and Delinquency Prevention, and the Office for Victims of Crime.

October 2009

Test Results for Forensic Media Preparation Tool:
Darik's Boot and Nuke 1.0.7

Contents

1. Results Summary 2
2. Test Case Selection 2
3. Test Materials 4
   3.1 Support Software 4
   3.2 Test Drive Creation 4
   3.3 Test Drive Analysis 4
   3.4 Test Drives 5
4. Test Results 5
   4.1 Test Results Report Key 5
   4.2 Test Details 6
      4.2.1 FMP-01-ATA28 6
      4.2.2 FMP-01-ATA48 8
      4.2.3 FMP-01-SATA28 9
      4.2.4 FMP-01-SATA48 11
      4.2.5 FMP-01-SCSI 12
      4.2.6 FMP-03-DCO 13
      4.2.7 FMP-03-DCO+HPA 15
      4.2.8 FMP-03-HPA 17

Introduction

The Computer Forensics Tool Testing (CFTT) program is a joint project of the National Institute of Justice (NIJ), the research and development organization of the U.S. Department of Justice (DOJ), and the National Institute of Standards and Technology's (NIST's) Office of Law Enforcement Standards and Information Technology Laboratory. CFTT is supported by other organizations, including the Federal Bureau of Investigation, the U.S. Department of Defense Cyber Crime Center, U.S. Internal Revenue Service Criminal Investigation Division Electronic Crimes Program, and the U.S. Department of Homeland Security's Bureau of Immigration and Customs Enforcement, U.S. Customs and Border Protection, and U.S. Secret Service. The objective of the CFTT program is to provide measurable assurance to practitioners, researchers, and other applicable users that the tools used in computer forensics investigations provide accurate results. Accomplishing this requires the development of specifications and test methods for computer forensics tools and subsequent testing of specific tools against those specifications.

Test results provide the information necessary for developers to improve tools, users to make informed choices, and the legal community and others to understand the tools' capabilities. The CFTT approach to testing computer forensic tools is based on well-recognized methodologies for conformance and quality testing. The specifications and test methods are posted on the CFTT Web site (http://www.cftt.nist.gov/) for review and comment by the computer forensics community.

This document reports the results from testing Darik's Boot and Nuke 1.0.7, against the *Forensic Media Preparation Tool Test Assertions and Test Plan Version 1.0*, available at the CFTT Web site (http://www.cftt.nist.gov/fmp-atp-pc-01.pdf).

Test results for other devices and software packages using the CFTT tool methodology can be found on NIJ's computer forensics tool testing Web page, http://www.ojp.usdoj.gov/nij/topics/technology/electronic-crime/cftt.htm.

Test Results for Forensic Media Preparation Tool

Tool Tested: **Darik's Boot and Nuke 1.0.7**
Version: **1.0.7**
Run Environments: Custom

Supplier: **Darik's Boot and Nuke**

Address: Vanadac Corporation
PO Box 660675 PMB 11493
Dallas, TX 75266–0675
United States

Tel: 1–866–969–3226
Email: support@dban.org
WWW: http://www.dban.org/

1. Results Summary

In all the test cases run against Darik's Boot and Nuke (DBAN) Version 1.0.7, all visible sectors were successfully overwritten. For the test cases that used drives containing an HPA or DCO, the tool behaved as designed by the vendor and did not overwrite hidden sectors.

- HPA remained intact, hidden sectors were not overwritten (FMP–03–HPA & FMP–03–DCO+HPA).
- DCO remained intact, hidden sectors were not overwritten (FMP–03–DCO & FMP–03–DCO+HPA).

2. Test Case Selection

Darik's Boot and Nuke software download version 1.0.7 was tested for its ability to overwrite sectors. The test cases selected were limited to only those test cases defined by *Forensic Media Preparation Tool Test Assertions and Test Plan Version 1.0* and applicable to features supported by this tool.

All selected test cases were *WRITE* tests (FMP–01 and FMP–03).

Three hidden sector test cases (FMP–03) were included among the cases selected. They were included to measure the tools' behavior in conjunction with hidden sectors. The tool documentation acknowledges that a drive may contain hidden sectors, but that the tool implementation leaves hidden sector content intact.

The following cases were used in testing Darik's Boot and Nuke 1.0.7:

- FMP–01–ATA28
- FMP–01–ATA48
- FMP–01–SATA28
- FMP–01–SATA48
- FMP–01–SCSI
- FMP–03–DCO
- FMP–03–DCO+HPA
- FMP–03–HPA

Since DBAN does not support a secure erase mode test cases FMP–02, FMP–04 and FMP–05 were omitted.

DBAN features an options menu from which a user can alter the test run behavior. Its options include:

- Wipe method
- PRNG (pseudo random number generator) schemes
- verification mode
- number of rounds

The available wipe methods for overwriting the visible sectors of a destination drive are the following:

- Quick Erase
- DoD 5220.22–M Short
- DoD 5220.22–M
- RCMP TSSIS OPS–II
- Guttman Wipe
- PRNG Stream

A test run was conducted by first selecting a wipe method from the options menu, then choosing additional parameters which controlled the length and depth of the run.

A note on verification mode: our testing methodology cannot detect if verification actually takes place or if the verification process can detect a failure to wipe.

The following source interfaces were tested: ATA28, ATA48, SATA28, SATA48 and SCSI.

3. Test Materials

3.1 Support Software

Several programs were used in the setup and analysis of the test drives. These include **hdat2** (download from http://www.hdat2.com/download.html), **dsumm** (download from http://www.cftt.nist.gov/), and **diskwipe** from **FS-TST Release 2.0** (download from http://www.cftt.nist.gov/diskimaging/fs-tst20.zip).

The **hdat2** program is used to create, remove and document hidden areas on a drive.

The **diskwipe** program initializes a hard drive with known content.

The **dsumm** program analyzes the content of a hard drive. It produces a summary of disk contents in terms of counts for each byte value present on the drive. For example, if a drive can contain 10GB (19531250 sectors of 512 bytes per sector) and the drive is wiped with zero bytes, then **dsumm** reports 10,000,000,000 zero bytes. The program also prints the first sector found with printable ASCII content.

3.2 Test Drive Creation

The following steps are used to setup a test drive:

1. The drive is initially filled with known content by the **diskwipe** program from FS-TST. The **diskwipe** program writes the sector address to each sector in both C/H/S and LBA format. The remainder of the sector bytes is set to a constant fill value unique for each drive. The fill value is noted in the **diskwipe** tool log file.
2. The drive content is analyzed by the **dsumm** program. This documents the content of the drive. Each sector has unique content after the setup.
3. If the drive is intended for hidden area tests (FMP–03), an HPA, a DCO or both are created.
4. The drive size after creation of a hidden area is recorded.

3.3 Test Drive Analysis

The following steps are used to analyze a test drive after it has been wiped by the tool under test:

1. The size of the drive is recorded. This determines if the tool changes the size of a hidden area.
2. Any hidden areas still present on the drive are removed.
3. The **dsumm** program is run to determine the final content of the drive.

3.4 Test Drives

The following hard drives were used in testing. The column labeled **Test Case** identifies the test case. The column labeled **Sectors** is the size of the drive with no DCO or HPA. The column labeled **Model** is the model of the drive as returned by the ATA IDENTIFY DEVICE command. The column labeled **Serial #** is the serial number as returned by the ATA IDENTIFY DEVICE command.

Test Case	Sectors	Model	Serial #
FMP-01-ATA28	156301488	WDC WD800BB-75CAA0	WD-WMA8E2108916
FMP-01-ATA48	488397168	WDC WD2500JB-00GVC0	WD-WCAL78188039
FMP-01-SATA28	78140160	FUJITSU MHW2040BH	K10XT7B278AP
FMP-01-SATA48	312581808	ST9160310AS	5SV092JK
FMP-01-SCSI	71721820	ATLAS10K2-TY367L	163022042046
FMP-03-DCO	488397168	WDC WD2500JB-00GVC0	WD-WCAL78188039
FMP-03-DCO+HPA	156301488	Hitachi HTS541680J9AT00	SB0241HGGAWY8E
FMP-03-HPA	78140160	FUJITSU MHW2040BH	K10XT7B278AP

For test cases FMP-03 the layout of visible and hidden sectors is as follows. The column labeled **Test Case** identifies the test case. The column labeled **Size** is the number of visible sectors presented to the device for the test case. The column labeled **Hidden** is the size of the hidden area.

Test Case	Size	Total	Hidden (DCO+HPA)
FMP-03-DCO	24419859	488397168	463977309
FMP-03-DCO+HPA	18756179	156301488	137545309
FMP-03-HPA	3907009	78140160	74233151

4. Test Results

The main item of interest for interpreting the test results is determining the conformance of the tool under test with the test assertions. Conformance with each assertion tested by a given test case is evaluated by examining the **Log Highlights** box of the test report summary.

4.1 Test Results Report Key

A summary of the actual test results is presented in this report. The following table presents a description of each section of the test report summary.

Heading	Description
First Line:	Test case ID, name, and version of tool tested.
Case Summary:	Test case summary from *Forensic Media Preparation Tool Test Assertions and Test Plan Version 1.0*.
Assertions:	The test assertions applicable to the test case, selected from *Forensic Media Preparation Tool Test Assertions and Test Plan Version 1.0*.

Heading	Description
Tester Name:	Name or initials of person executing test procedure.
Analysis Host:	Host used to setup test drive and analyze final drive state.
Test Host:	Host computer executing the test.
Test Date:	Time and date that test was started.
Test Drive:	Drive erased by the tool under test.
Source Setup:	Report of the native drive size, the size of any hidden areas, the apparent size of the drive (as reported by an ATA IDENTIFY DEVICE command) and an analysis of initial drive contents.
Tool Settings:	Selected wipe options: method,: PRNG, verify, rounds.
Log Highlights:	Report of the state of the drive after executing the tool under test, including the apparent drive size, size of hidden area and analysis of drive contents. The ASCII content of the first nonbinary-zero sector is reported.
Results:	Expected and actual results for each assertion tested.
Analysis:	Whether or not the expected results were achieved.

4.2 Test Details

4.2.1 FMP-01-ATA28

Test Case FMP-01-ATA28 Darik's Boot and Nuke 1.0.7	
Case Summary:	FMP-01. Overwrite visible sectors using WRITE commands.
Assertions:	FMP-CA-01 All visible sectors shall be overwritten with the specified benign data.
Tester Name:	csr
Analysis host:	frank
Test host:	frank
Test date:	Mon Jun 8 16:03:45 2009
Test drive:	56-IDE
Source Setup:	Initial setup size: 156301488 from total of 156301488 (with 0 hidden) IDE disk: Model (WDC WD800BB-75CAA0) serial # (WD-WMA8E2108916) Sector 0 is first sector with printable text ============= Start text ============= 00000/000/01 000000000000VVVVVVVVVVVVVVVVVVVVVVVVVVVVVVVV VVVVVVVVVVVVVVVVVVVVVVVVVVVVVVVVVVVVVVVVVVVVVVVVVVVVVVVV VVVVVVVVVVVVVVVVVVVVVVVVVVVVVVVVVVVVVVVVVVVVVVVVVVVVVVVV VVVVVVVVVVVVVVVVVVVVVVVVVVVVVVVVVVVVVVVVVVVVVVVVVVVVVVVV VVVVVVVVVVVVVVVVVVVVVVVVVVVVVVVVVVVVVVVVVVVVVVVVVVVVVVVV VVVVVVVVVVVVVVVVVVVVVVVVVVVVVVVVVVVVVVVVVVVVVVVVVVVVVVVV VVVVVVVVVVVVVVVVVVVVVVVVVVVVVVVVVVVVVVVVVVVVVVVVVVVVVVVV VVVVVVVVVVVVVVVVVVVVVVVVVVVVVVVVVVVVVVVVVVVVVVVVVVVVVVVV VVVVVVVVVVVVVVVVVVVVVVVVVVV ============= End text Sector 0 ============= 9 <new line> characters inserted for readability Totals for all sectors summary format: <count> <hex value> <(actual character if printable)> ... 156301488 00 156301488 20 ( ) 312602976 2F (/) 1092738319 30 (0) 445157427 31 (1) 274740905 32 (2) 274642393 33 (3) 272159917 34 (4) 262536293 35 (5) 225709546 36 (6) 215483146 37 (7) 215483143 38 (8) 215483135 39 (9) 75907021680 56 (V)

Test Case FMP-01-ATA28 Darik's Boot and Nuke 1.0.7	
	Totals for non-ASCII sectors summary format: <count> <hex value> <(actual character if printable)> ... 80026361856 bytes, 156301488 sectors, 14 distinct values seen 156301488 sectors have printable text
Tool Settings:	Method: DoD Short PRNG: Mersenne Twister Verify: Off Rounds: Default
Log Highlights:	Size after tool runs: 156301488 from total of 156301488 (with 0 hidden) Analysis of tool result -- Totals for all sectors summary format: <count> <hex value> <(actual character if printable)> ... 80026361856 00 Totals for non-ASCII sectors summary format: <count> <hex value> <(actual character if printable)> ... 80026361856 00 80026361856 bytes, 156301488 sectors, 1 distinct values seen No sectors have printable text
Results:	**Assertion & Expected Result** / **Actual Result** FMP-CA-01 Visible sectors overwritten / as expected
Analysis:	Expected results achieved

4.2.2 FMP-01-ATA48

Test Case FMP-01-ATA48 Darik's Boot and Nuke 1.0.7	
Case Summary:	FMP-01. Overwrite visible sectors using WRITE commands.
Assertions:	FMP-CA-01 All visible sectors shall be overwritten with the specified benign data.
Tester Name:	csr
Analysis host:	frank
Test host:	frank
Test date:	Wed Jun 10 08:25:15 2009
Test drive:	29-IDE

Source Setup:	Initial setup size: 488397168 from total of 488397168 (with 0 hidden)
	IDE disk: Model (WDC WD2500JB-00GVC0) serial # (WD-WCAL78188039)
	
	Sector 0 is first sector with printable text
	============= Start text =============
	00000/000/01 000000000000))))))))))))))))))))))))))))))))
	))
	))
	))
	))
	))
	))
	))
	))))))))))))))))))))))))))))))))))
	============= End text Sector 0 =============
	9 <new line> characters inserted for readability
	
	Totals for all sectors
	summary format: <count> <hex value> <(actual character if printable)> ...
	  488397168 00 488397168 20 ( ) 237361023648 29 ()) 976794336 2F (/)
	  2735169210 30 (0) 1278997882 31 (1) 1192805876 32 (2) 933260747 33 (3)
	  905775911 34 (4) 805865997 35 (5) 749775664 36 (6) 718765480 37 (7)
	  716559080 38 (8) 707761849 39 (9)
	Totals for non-ASCII sectors
	summary format: <count> <hex value> <(actual character if printable)> ...
	
	250059350016 bytes, 488397168 sectors, 14 distinct values seen
	488397168 sectors have printable text

Tool Settings:	Method: Quick Erase PRNG: Issac Verify: Last Rounds: 2
Log Highlights:	Size after tool runs: 488397168 from total of 488397168 (with 0 hidden) Analysis of tool result -- Totals for all sectors summary format: <count> <hex value> <(actual character if printable)> ... 250059350016 00 Totals for non-ASCII sectors summary format: <count> <hex value> <(actual character if printable)> ... 250059350016 00 250059350016 bytes, 488397168 sectors, 1 distinct values seen No sectors have printable text

Results:	Assertion & Expected Result	Actual Result
	FMP-CA-01 Visible sectors overwritten	as expected
Analysis:	Expected results achieved	

4.2.3 FMP-01-SATA28

Test Case FMP-01-SATA28 Darik's Boot and Nuke 1.0.7	
Case Summary:	FMP-01. Overwrite visible sectors using WRITE commands.
Assertions:	FMP-CA-01 All visible sectors shall be overwritten with the specified benign data.
Tester Name:	csr
Analysis host:	frank
Test host:	frank
Test date:	Thu Jun 11 10:21:22 2009
Test drive:	24-LAP

Source Setup:	Initial setup size: 78140160 from total of 78140160 (with 0 hidden)
	IDE disk: Model (FUJITSU MHW2040BH) serial # (K10XT7B278AP)
	
	Sector 0 is first sector with printable text
	============= Start text =============
	00000/000/01 000000000000$$$$$$$$$$$$$$$$$$$$$$$$$$$$$$$$$$
	$$
	$$
	$$
	$$
	$$
	$$
	$$$$$$$$$$$$$$$$$$$$$$$$$$$$$$$
	============= End text Sector 0 =============
	9 <new line> characters inserted for readability
	
	Totals for all sectors
	summary format: <count> <hex value> <(actual character if printable)> ...
	  78140160 00 78140160 20 ( ) 37976117760 24 ($) 156280320 2F (/)
	 561878293 30 (0) 173598093 31 (1) 159768433 32 (2) 142914673 33 (3)
	 139463608 34 (4) 123744696 35 (5) 114674216 36 (6) 107788836 37 (7)
	  98210496 38 (8) 97042176 39 (9)
	Totals for non-ASCII sectors
	summary format: <count> <hex value> <(actual character if printable)> ...
	
	40007761920 bytes, 78140160 sectors, 14 distinct values seen
	78140160 sectors have printable text

Tool Settings:	Method: RCMP TSSIT OPS-II PRNG: Issac Verfiy: Each Rounds: 1
Log Highlights:	Size after tool runs: 78140160 from total of 78140160 (with 0 hidden) Analysis of tool result -- Sector 0 is first sector with printable text ============= Start text ============= Ka4lu\qt|m{u\$YNM~@?LClXY1`\V{,8BNB(+TPw\GA&El)@G%H/DAmJ:{0\$Z [*OG&{!eE{EDQb#QQ37Y~jR02O\h|SwmZ"\_Q*,`Mb"[=^l3x=-}CN7D+_?No s9XT~CQHO9N7qrK>:,HL_;/,CoZTzVIe.Up;H,;me.v)h:2Ead[AIh=6 L.8 P,ao"j4/Kzs%I=1c*?* ============= End text Sector 0 ============= 4 <new line> characters inserted for readability Totals for all sectors summary format: <count> <hex value> <(actual character if printable)> ... 156254893 00 156299920 01 156286646 02 156275527 03 156293076 04 156290614 05 156289403 06 156284130 07 156292799 08 156271928 09 156291261 0A 156275711 0B 156268407 0C 156282389 0D 156287707 0E 156268326 0F 156278915 10 156268392 11 156292161 12 156259198 13 156288623 14 156274040 15 156284860 16 156288128 17 156249780 18 156275366 19 156288661 1A 156288715 1B 156259755 1C 156257631 1D 156271960 1E 156280415 1F

Test Case FMP-01-SATA28 Darik's Boot and Nuke 1.0.7
156265700 20 ( )   156264777 21 (!)   156272114 22 (")   156280545 23 (#)
156277942 24 ($)   156290069 25 (%)   156294838 26 (&)   156275629 27 (')
156289816 28 (()   156301590 29 ())   156267521 2A (*)   156280475 2B (+)
156276292 2C (,)   156269278 2D (-)   156266320 2E (.)   156263908 2F (/)
156283948 30 (0)   156286228 31 (1)   156272161 32 (2)   156273640 33 (3)
156283157 34 (4)   156296786 35 (5)   156289135 36 (6)   156272958 37 (7)
156310472 38 (8)   156270965 39 (9)   156283348 3A (:)   156277440 3B (;)
156287908 3C (<)   156259095 3D (=)   156277656 3E (>)   156274600 3F (?)
156295306 40 (@)   156267464 41 (A)   156268463 42 (B)   156297526 43 (C)
156286039 44 (D)   156270271 45 (E)   156269838 46 (F)   156282091 47 (G)
156282765 48 (H)   156273927 49 (I)   156289788 4A (J)   156277177 4B (K)
156294918 4C (L)   156291059 4D (M)   156257881 4E (N)   156276620 4F (O)
156280706 50 (P)   156273157 51 (Q)   156282564 52 (R)   156270985 53 (S)
156280018 54 (T)   156284799 55 (U)   156284811 56 (V)   156278781 57 (W)
156275939 58 (X)   156271612 59 (Y)   156284489 5A (Z)   156289751 5B ([)
156268768 5C (\)   156285682 5D (])   156266022 5E (^)   156287220 5F (_)
156282295 60 (`)   156269572 61 (a)   156285984 62 (b)   156284115 63 (c)
156297470 64 (d)   156286337 65 (e)   156298271 66 (f)   156277352 67 (g)
156282912 68 (h)   156265934 69 (i)   156299716 6A (j)   156259814 6B (k)
156288044 6C (l)   156272768 6D (m)   156288398 6E (n)   156279853 6F (o)
156276906 70 (p)   156279836 71 (q)   156262658 72 (r)   156270404 73 (s)
156283598 74 (t)   156276615 75 (u)   156267416 76 (v)   156283893 77 (w)
156304627 78 (x)   156286077 79 (y)   156259167 7A (z)   156277438 7B ({)
156295232 7C (|)   156269730 7D (})   156312324 7E (~)   156284524 7F
156276520 80   156273240 81   156286601 82   156285003 83
156297827 84   156268871 85   156259966 86   156295815 87
156301121 88   156262895 89   156272131 8A   156287971 8B
156292588 8C   156268899 8D   156299772 8E   156308323 8F
156281860 90   156276681 91   156267198 92   156295375 93
156271944 94   156277474 95   156306842 96   156269589 97
156276301 98   156272065 99   156301085 9A   156292547 9B
156300263 9C   156287388 9D   156280600 9E   156287180 9F
156264533 A0   156257819 A1   156278466 A2   156273435 A3
156263302 A4   156291291 A5   156279360 A6   156268677 A7
156272230 A8   156278478 A9   156288166 AA   156273191 AB
156294495 AC   156275514 AD   156288618 AE   156276492 AF
156259571 B0   156287790 B1   156291043 B2   156233538 B3
156265272 B4   156285531 B5   156278565 B6   156296510 B7
156267502 B8   156291093 B9   156289856 BA   156277048 BB
156287020 BC   156261271 BD   156277867 BE   156278196 BF
156300200 C0   156285916 C1   156309929 C2   156289265 C3
156277435 C4   156265953 C5   156263501 C6   156262474 C7
156289045 C8   156278338 C9   156279572 CA   156291648 CB
156271208 CC   156283680 CD   156295155 CE   156287752 CF
156281234 D0   156291897 D1   156270577 D2   156297664 D3
156279095 D4   156282326 D5   156289348 D6   156271537 D7
156277371 D8   156267525 D9   156285857 DA   156267705 DB
156267351 DC   156303317 DD   156275820 DE   156285552 DF
156275852 E0   156314749 E1   156270426 E2   156294172 E3
156284601 E4   156272469 E5   156300845 E6   156287724 E7
156269592 E8   156322495 E9   156274885 EA   156282922 EB
156270865 EC   156270812 ED   156261325 EE   156298407 EF
156290556 F0   156262807 F1   156284291 F2   156247612 F3
156272562 F4   156280839 F5   156303690 F6   156275273 F7
156272033 F8   156286894 F9   156283047 FA   156283442 FB
156279917 FC   156257461 FD   156289933 FE   156269135 FF
Totals for non-ASCII sectors
summary format: <count> <hex value> <(actual character if printable)> ...

40007761920 bytes, 78140160 sectors, 256 distinct values seen
78140160 sectors have printable text |

Results:	Assertion & Expected Result	Actual Result
	FMP-CA-01 Visible sectors overwritten	as expected
Analysis:	Expected results achieved	

4.2.4 FMP-01-SATA48

Test Case FMP-01-SATA48 Darik's Boot and Nuke 1.0.7	
Case Summary:	FMP-01. Overwrite visible sectors using WRITE commands.
Assertions:	FMP-CA-01 All visible sectors shall be overwritten with the specified benign data.
Tester Name:	csr
Analysis host:	frank
Test host:	frank
Test date:	Fri Jun 12 12:41:02 2009
Test drive:	21-LAP

Source Setup:	Initial setup size: 312581808 from total of 312581808 (with 0 hidden)
	IDE disk: Model (ST9160310AS) serial # (5SV092JK)
	
	Sector 0 is first sector with printable text
	============= Start text =============
	00000/000/01 000000000000!!!!!!!!!!!!!!!!!!!!!!!!!!!!!!!!!!
	!!
	!!
	!!
	!!
	!!
	!!
	!!!!!!!!!!!!!!!!!!!!!!!!!!!!!!!
	============= End text Sector 0 =============
	9 <new line> characters inserted for readability
	
	Totals for all sectors
	summary format: <count> <hex value> <(actual character if printable)> ...
	  312581808 00 312581808 20 ( ) 151914758688 21 (!) 625163616 2F (/)
	 1850492169 30 (0) 906528227 31 (1) 696435016 32 (2) 541016511 33 (3)
	  522787395 34 (4) 514450557 35 (5) 478352540 36 (6) 458495114 37 (7)
	  458481159 38 (8) 449761088 39 (9)
	Totals for non-ASCII sectors
	summary format: <count> <hex value> <(actual character if printable)> ...
	
	160041885696 bytes, 312581808 sectors, 14 distinct values seen
	312581808 sectors have printable text

Tool Settings:	Method: Quick Erase PRNG: Mersenne Twister Verify: Off Rounds: 1
Log Highlights:	Size after tool runs: 312581808 from total of 312581808 (with 0 hidden) Analysis of tool result -- Totals for all sectors summary format: <count> <hex value> <(actual character if printable)> ... 160041885696 00 Totals for non-ASCII sectors summary format: <count> <hex value> <(actual character if printable)> ... 160041885696 00 160041885696 bytes, 312581808 sectors, 1 distinct values seen No sectors have printable text
Results:	**Assertion & Expected Result** **Actual Result** FMP-CA-01 Visible sectors overwritten as expected
Analysis:	Expected results achieved

4.2.5 FMP-01-SCSI

Test Case FMP-01-SCSI Darik's Boot and Nuke 1.0.7	
Case Summary:	FMP-01. Overwrite visible sectors using WRITE commands.
Assertions:	FMP-CA-01 All visible sectors shall be overwritten with the specified benign data.
Tester Name:	csr
Analysis host:	frank
Test host:	frank
Test date:	Tue Jun 23 13:14:35 2009
Test drive:	06

Source Setup:	Initial setup size: 71721820 from total of 71721820 (with 0 hidden)
	Model (ATLAS10K2-TY367L) serial # (163022042046)
	
	Sector 0 is first sector with printable text
	============= Start text =============
	00000/000/01 000000000000
	============= End text Sector 0 =============
	1 \<new line\> character inserted for readability
	
	Totals for all sectors
	summary format: \<count\> \<hex value\> \<(actual character if printable)\> ...
	  71721820 00 34856804520 06 71721820 20 ( ) 143443640 2F (/)
	  519143675 30 (0) 162528133 31 (1) 149139936 32 (2) 133670254 33 (3)
	  123349540 34 (4) 113156848 35 (5) 104831312 36 (6) 91849268 37 (7)
	  90105547 38 (8) 90105527 39 (9)
	Totals for non-ASCII sectors
	summary format: \<count\> \<hex value\> \<(actual character if printable)\> ...
	
	36721571840 bytes, 71721820 sectors, 14 distinct values seen
	71721820 sectors have printable text

Tool Settings:	Method: RCMP TSSIT OPS-II PRNG: Issac Verify: Each Rounds: 1
Log Highlights:	Size after tool runs: 71721820 from total of 71721820 (with 0 hidden) Analysis of tool result -- Totals for all sectors summary format: \<count\> \<hex value\> \<(actual character if printable)\> ... 36721571840 00 Totals for non-ASCII sectors summary format: \<count\> \<hex value\> \<(actual character if printable)\> ... 36721571840 00 36721571840 bytes, 71721820 sectors, 1 distinct values seen No sectors have printable text
Results:	**Assertion & Expected Result** | **Actual Result** FMP-CA-01 Visible sectors overwritten | as expected
Analysis:	Expected results achieved

4.2.6 FMP-03-DCO

Test Case FMP-03-DCO Darik's Boot and Nuke 1.0.7	
Case Summary:	FMP-03. Overwrite hidden sectors using WRITE commands.
Assertions:	FMP-CA-01 All visible sectors shall be overwritten with the specified benign data. FMP-AO-01 If there is a hidden area present and the tool supports overwriting sectors contained in a hidden area, then all sectors contained in the hidden area shall be overwritten with the specified benign data. FMP-AO-02 A hidden area may optionally be removed from the storage device.
Tester Name:	csr
Analysis host:	frank
Test host:	frank
Test date:	Tue Jun 16 08:34:04 2009
Test drive:	29-IDE

Source Setup:	Initial setup size: 24419859 from total of 488397168 (with 463977309 hidden)
	IDE disk: Model (WDC WD2500JB-00GVC0) serial # (WD-WCAL78188039)
	
	Sector 0 is first sector with printable text
	============= Start text =============
	00000/000/01 000000000000))))))))))))))))))))))))))))))))))))
	))
	))
	))
	))
	))
	))
	))
	))))))))))))))))))))))))))))))
	============= End text Sector 0 =============
	9 <new line> characters inserted for readability
	
	Totals for all sectors
	summary format: <count> <hex value> <(actual character if printable)> ...
	  24419859 00 24419859 20 ( ) 11868051474 29 ()) 48839718 2F (/)
	 194863239 30 (0) 65826577 31 (1) 47412497 32 (2) 37724779 33 (3)
	  36676881 34 (4) 34378453 35 (5) 31251575 36 (6) 29701060 37 (7)
	  29701019 38 (8) 29700818 39 (9)
	Totals for non-ASCII sectors
	summary format: <count> <hex value> <(actual character if printable)> ...
	
	12502967808 bytes, 24419859 sectors, 14 distinct values seen
	24419859 sectors have printable text

Tool Settings:	Method: DoD 5220.22-M PRNG: Issac Verify: Off Rounds: 2
Log Highlights:	Size after tool runs: 24419859 from total of 488397168 (with 463977309 hidden) Analysis of tool result -- Sector 24419859 is first sector with printable text ============= Start text ============= 01520/016/52 000024419859)))))))))))))))))))))))))))))))))))))))))))))))))))))))))))))))))))))))))))))))))) ============= End text Sector 24419859 ============= 9 <new line> characters inserted for readability Totals for all sectors

Test Case FMP-03-DCO Darik's Boot and Nuke 1.0.7			

	summary format: <count> <hex value> <(actual character if printable)> ...
	 2966945117 00 463977309 20 ( ) 225492972174 29 ()) 927954618 2F (/)
	 2540305971 30 (0) 1213171305 31 (1) 1145393379 32 (2) 895535968 33 (3)
	  869099030 34 (4) 771487544 35 (5) 718524089 36 (6) 689064420 37 (7)
	  686858061 38 (8) 678061031 39 (9)
	 Totals for non-ASCII sectors
	 summary format: <count> <hex value> <(actual character if printable)> ...
	 2502967808 00
	
	 250059350016 bytes, 488397168 sectors, 14 distinct values seen
	 463977309 sectors have printable text			

Results:	**Assertion & Expected Result**		**Actual Result**
	FMP-CA-01 Visible sectors overwritten		as expected
	FMP-AO-01 Hidden sectors overwritten		DCO not overwritten
	FMP-AO-02 Hidden area final state is		in place
Analysis:	Expected results not achieved		

4.2.7 FMP-03-DCO+HPA

Test Case FMP-03-DCO+HPA Darik's Boot and Nuke 1.0.7	
Case Summary:	FMP-03. Overwrite hidden sectors using WRITE commands.
Assertions:	FMP-CA-01 All visible sectors shall be overwritten with the specified benign data. FMP-AO-01 If there is a hidden area present and the tool supports overwriting sectors contained in a hidden area, then all sectors contained in the hidden area shall be overwritten with the specified benign data. FMP-AO-02 A hidden area may optionally be removed from the storage device.
Tester Name:	csr
Analysis host:	frank
Test host:	frank
Test date:	Thu Jun 18 16:01:21 2009
Test drive:	15-LAP

Source Setup:	Initial setup size: 18756179 from total of 156301488 (with 137545309 hidden)
	IDE disk: Model (Hitachi HTS541680J9AT00) serial # (SB0241HGGAWY8E)
	
	Sector 0 is first sector with printable text
	============= Start text =============
	00000/000/01 000000000000
	============= End text Sector 0 =============
	1 <new line> character inserted for readability
	
	Totals for all sectors
	summary format: <count> <hex value> <(actual character if printable)> ...
	  23445223 00 11394378378 15 23445223 20 ( ) 46890446 2F (/)
	 188316972 30 (0) 63144036 31 (1) 45072570 32 (2) 36017102 33 (3)
	  34487902 34 (4) 32921277 35 (5) 30077619 36 (6) 28589035 37 (7)
	  28589035 38 (8) 28579358 39 (9)
	Totals for non-ASCII sectors
	summary format: <count> <hex value> <(actual character if printable)> ...
	
	12003954176 bytes, 23445223 sectors, 14 distinct values seen
	23445223 sectors have printable text

Tool Settings:	Method: PRNG Stream PRNG: Issac Verify: Each Rounds: 1

Log Highlights:	Size after tool runs: 18756179 from total of 156301488 (with 137545309 hidden)
	Analysis of tool result --
	
	Sector 18756179 is first sector with printable text
	============= Start text =============
	01167/132/09 000018756179
	============= End text Sector 18756179 =============
	1 <new line> character inserted for readability
	
	Totals for all sectors
	summary format: <count> <hex value> <(actual character if printable)> ...
	 9740708957 00 66847020174 15 137545309 20 ( ) 275090618 2F (/)
	  895308717 30 (0) 339688267 31 (1) 271636254 32 (2) 241721228 33 (3)
	  239536603 34 (4) 232605388 35 (5) 209855675 36 (6) 199858160 37 (7)
	  199694935 38 (8) 196091571 39 (9)
	Totals for non-ASCII sectors
	summary format: <count> <hex value> <(actual character if printable)> ...
	9603163648 00
	
	80026361856 bytes, 156301488 sectors, 14 distinct values seen
	137545309 sectors have printable text

Results:	Assertion & Expected Result	Actual Result	
	FMP-CA-01 Visible sectors overwritten	as expected	
	FMP-AO-01 Hidden sectors overwritten	DCO+HPA not overwritten	

Test Case FMP-03-DCO+HPA Darik's Boot and Nuke 1.0.7			
	FMP-AO-02 Hidden area final state is	in place	
Analysis:	Expected results not achieved		

4.2.8 FMP-03-HPA

Test Case FMP-03-HPA Darik's Boot and Nuke 1.0.7	
Case Summary:	FMP-03. Overwrite hidden sectors using WRITE commands.
Assertions:	FMP-CA-01 All visible sectors shall be overwritten with the specified benign data. FMP-AO-01 If there is a hidden area present and the tool supports overwriting sectors contained in a hidden area, then all sectors contained in the hidden area shall be overwritten with the specified benign data. FMP-AO-02 A hidden area may optionally be removed from the storage device.
Tester Name:	csr
Analysis host:	frank
Test host:	frank
Test date:	Wed Jun 17 08:39:15 2009
Test drive:	24-LAP

Source Setup:	Initial setup size: 3907009 from total of 78140160 (with 74233151 hidden)
	IDE disk: Model (FUJITSU MHW2040BH) serial # (K10XT7B278AP)
	
	Sector 0 is first sector with printable text
	============= Start text =============
	00000/000/01 000000000000$$$$$$$$$$$$$$$$$$$$$$$$$$$$$$$$$$$$
	$$
	$$
	$$
	$$
	$$
	$$
	$$
	$$$$$$$$$$$$$$$$$$$$$$$$$$$$$$$$$
	============= End text Sector 0 =============
	9 <new line> characters inserted for readability
	
	Totals for all sectors
	summary format: <count> <hex value> <(actual character if printable)> ...
	  78140160 00 78140160 20 ( ) 37976117760 24 ($) 156280320 2F (/)
	  561878293 30 (0) 173598093 31 (1) 159768433 32 (2) 142914673 33 (3)
	  139463608 34 (4) 123744696 35 (5) 114674216 36 (6) 107788836 37 (7)
	  98210496 38 (8) 97042176 39 (9)
	Totals for non-ASCII sectors
	summary format: <count> <hex value> <(actual character if printable)> ...
	
	40007761920 bytes, 78140160 sectors, 14 distinct values seen
	78140160 sectors have printable text

Tool Settings:	Method: Guttman PRNG: Mersenne Twister Verify: Off Rounds: 1
Log Highlights:	Size after tool runs: 3907009 from total of 78140160 (with 74233151 hidden) Analysis of tool result -- Sector 3907009 is first sector with printable text ============= Start text ============= 00243/051/02 000003907009$$$$$$$$$$$$$$$$$$$$$$$$$$$$$$$$$$$$ $$ $$ $$ $$ $$ $$ $$ $$$$$$$$$$$$$$$$$$$$$$$$$$$$$$$$$ ============= End text Sector 3907009 ============= 9 <new line> characters inserted for readability Totals for all sectors

Test Case FMP-03-HPA Darik's Boot and Nuke 1.0.7			
	summary format: <count> <hex value> <(actual character if printable)> ... 2074621759 00 74233151 20 ( ) 36077311386 24 ($) 148466302 2F (/) 525356360 30 (0) 164311494 31 (1) 152083232 32 (2) 136871318 33 (3) 134501733 34 (4) 118926650 35 (5) 110304875 36 (6) 103668550 37 (7) 94090219 38 (8) 93014891 39 (9) Totals for non-ASCII sectors summary format: <count> <hex value> <(actual character if printable)> ... 2000388608 00 40007761920 bytes, 78140160 sectors, 14 distinct values seen 74233151 sectors have printable text		
Results:	**Assertion & Expected Result**	**Actual Result**	
	FMP-CA-01 Visible sectors overwritten	as expected	
	FMP-AO-01 Hidden sectors overwritten	HPA not overwritten	
	FMP-AO-02 Hidden area final state is	in place	
Analysis:	Expected results not achieved		

About the National Institute of Justice

NIJ is the research, development, and evaluation agency of the U.S. Department of Justice. NIJ's mission is to advance scientific research, development, and evaluation to enhance the administration of justice and public safety. NIJ's principal authorities are derived from the Omnibus Crime Control and Safe Streets Act of 1968, as amended (see 42 U.S.C. §§ 3721–3723).

The NIJ Director is appointed by the President and confirmed by the Senate. The Director establishes the Institute's objectives, guided by the priorities of the Office of Justice Programs, the U.S. Department of Justice, and the needs of the field. The Institute actively solicits the views of criminal justice and other professionals and researchers to inform its search for the knowledge and tools to guide policy and practice.

Strategic Goals

NIJ has seven strategic goals grouped into three categories:

Creating relevant knowledge and tools

1. Partner with State and local practitioners and policymakers to identify social science research and technology needs.
2. Create scientific, relevant, and reliable knowledge—with a particular emphasis on terrorism, violent crime, drugs and crime, cost-effectiveness, and community-based efforts—to enhance the administration of justice and public safety.
3. Develop affordable and effective tools and technologies to enhance the administration of justice and public safety.

Dissemination

4. Disseminate relevant knowledge and information to practitioners and policymakers in an understandable, timely, and concise manner.
5. Act as an honest broker to identify the information, tools, and technologies that respond to the needs of stakeholders.

Agency management

6. Practice fairness and openness in the research and development process.
7. Ensure professionalism, excellence, accountability, cost-effectiveness, and integrity in the management and conduct of NIJ activities and programs.

Program Areas

In addressing these strategic challenges, the Institute is involved in the following program areas: crime control and prevention, including policing; drugs and crime; justice systems and offender behavior, including corrections; violence and victimization; communications and information technologies; critical incident response; investigative and forensic sciences, including DNA; less-than-lethal technologies; officer protection; education and training technologies; testing and standards; technology assistance to law enforcement and corrections agencies; field testing of promising programs; and international crime control.

In addition to sponsoring research and development and technology assistance, NIJ evaluates programs, policies, and technologies. NIJ communicates its research and evaluation findings through conferences and print and electronic media.

To find out more about the National Institute of Justice, please visit:

http://www.ojp.usdoj.gov/nij

or contact:

National Criminal Justice
 Reference Service
P.O. Box 6000
Rockville, MD 20849–6000
800–851–3420
http://www.ncjrs.gov

www.ingramcontent.com/pod-product-compliance
Lightning Source LLC
Chambersburg PA
CBHW081824170526
45167CB00008B/3536